BLASTOFF!

URANUS

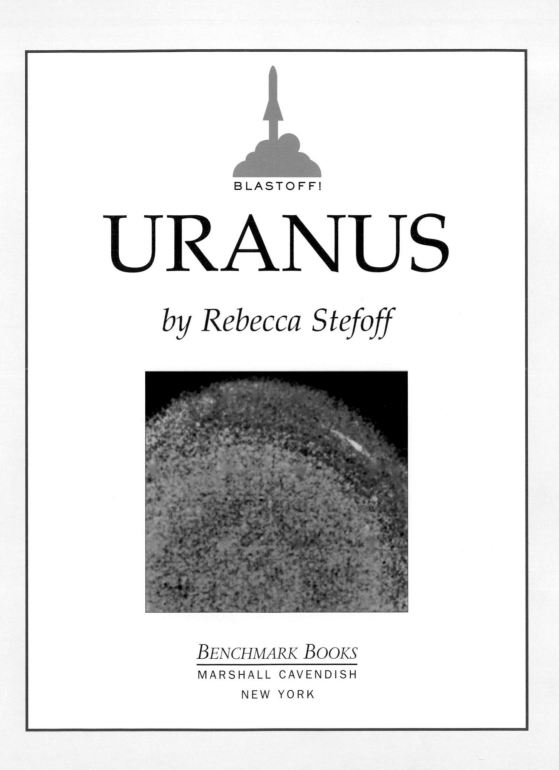

BLASTOFF!

URANUS

by Rebecca Stefoff

BENCHMARK BOOKS

MARSHALL CAVENDISH

NEW YORK

With special thanks to Professor Jerry LaSala, University of Maine,
for his careful review of the manuscript.

Benchmark Books
Marshall Cavendish
99 White Plains Road
Tarrytown, NY 10591
www.marshallcavendish.com

Library of Congress Cataloging-in-Publication Data

Stefoff, Rebecca, 1951-
Uranus / by Rebecca Stefoff.
p. cm. — (Blastoff!)
Includes bibliographical references and index.
Summary: Describes how Uranus, the seventh planet from the Sun, was first discovered,
how it was studied in the twentieth century, and what we know about it today.
ISBN 0-7614-1401-0
1. Uranus (Planet)—Juvenile literature. [1. Uranus {Planet}] I. Title. II. Series.
QB681 .S74 2002   523.47-dc21   2001043805

Printed in Italy
1 3 5 6 4 2

Photo Research by Anne Burns Images
Cover Photo: Photri, Inc.

The photographs in this book are used by permission and through the courtesy of: Bridgeman
Art Library, 8, 55; The Stapleton Collection, 14; Private Collection Photo Researchers, 10, 56;
Science Photo Library, 11, 23, 45; David Hardy/Science Photo Library, 13; Royal Observatory,
Edinburgh/SPL, 26, 34; Julian Baum/Science Photo Library, 27; Chris Butler/Science Photo
Library, 32; Sally Benusen/Science Photo Library (1987), 42; U.S. Geological Society/Science
Photo Library, 51; Mark Garlick, Science Photo Library, 31; Space Telescope Science
Institute/NASA/Science Photo Library Photri-Microstock, 17, 40, 46, 48 (upper left); NASA, 7,
21, 24, 37, 39, 52; Astronomical Society of the Pacific, 41; JPL, 48 (lower right), 57.

Cover: A computer-generated image of Uranus based on data from Voyager 2 *(the circles
were created during the imaging process and do not really appear on the planet).*

CONTENTS

1

THE SURPRISING
SEVENTH PLANET

In the outer reaches of the Solar System, far from the Sun's light and warmth, a cloud-wrapped world rolls slowly through space. It is Uranus (pronounced YOOR anniss), the seventh planet from the Sun. Many things about Uranus have surprised astronomers, but most surprising of all was the simple fact of its existence. Uranus was the first planet found in modern times, and its discovery was astonishing. Like a rock crashing through a window of colored glass, Uranus splintered ancient ideas about the heavens and opened a new view onto the Solar System.

THE SUN, THE MOON, FIVE WANDERERS, AND EARTH

People have gazed at the night sky for thousands of years. Records left by early civilizations such as the Babylonians in Iraq and the Maya in Central America show that many ancient peoples were very familiar with the motions of the stars. They saw how the stars form a great pattern that appears to wheel across the heavens over the course of a night and changes with the seasons. They also knew the seasonal appearances of the biggest and brightest objects in the

This view of Uranus shows an artist's view of the rings around the planet discovered by a team of astronomers who were gathering data on the planet's diameter. This was the first major structural find in the Solar System since the discovery of Pluto in 1930.

6 URANUS

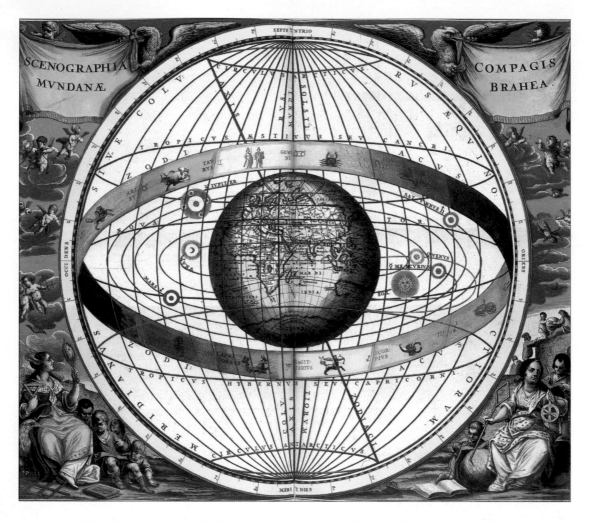

Published in an atlas of the heavens in 1660–1661, this map shows the old geocentric, or Earth-centered, view of the Solar System. The Sun and planets move around Earth in complex orbits.

sky, the Sun and Moon, and made accurate calendars based on their movements and positions.

Five points of light in the night sky were different from the stars and the Moon. They were small, like the stars, but each seemed set on its own course, moving independently of the others. Although the stars always remained in the same positions relative to one another, these five celestial objects slowly changed their positions against the starry backdrop. The ancient Greeks named them *planetes asters*, or "wandering stars." They came to be known as the planets, with names from Greek and Roman mythology: Mercury, messenger of the gods; Venus, goddess of love; Mars, god of war; Jupiter, ruler of the gods; and Saturn, father of Jupiter.

The most widely held view of the heavens was that the Sun, the Moon, and the five planets revolved around Earth, which lay at the center of the Universe. During the sixteenth century, however, a handful of European scientists challenged that geocentric view. In 1543, Polish astronomer Nicolaus Copernicus published his theory that the planets revolve around the Sun, not around Earth. A few years later, Danish astronomer Tycho Brahe made a very detailed survey of the night sky. Brahe thought the planets revolved around the Sun and that the Sun, with the planets, revolved around Earth. But German mathematician Johannes Kepler used Brahe's observations to show that Copernicus had been right. Kepler proved that Earth is a planet like the other five, revolving around the Sun. This new understanding of the heavens upset people who preferred to believe that God had placed Earth at the center of the Universe.

At times, knowledge is dangerous to those who possess it. The Roman Catholic Church punished scientists and thinkers—with death, in some cases—for writing that Earth and the other planets revolve around the Sun. But although the idea of a Sun-centered planetary system was unwelcome to some people, it gained wider acceptance after Italian scientist Galileo Galilei made telescopes that allowed closer study of the heavens.

By the eighteenth century most educated people accepted the fact that Earth is a planet and that the planets revolve around the Sun.

Polish astronomer Nicolaus Copernicus decided that Earth and the other planets revolve around the Sun, but delayed publishing his findings, fearing hostility from church officials. Indeed, the Roman Catholic Church banned his book until 1835, nearly three centuries after his death.

In fact, they thought they now understood the Solar System perfectly. The Solar System consisted of the Sun and six planets, and it ended at the orbit of Saturn. But this view of the Solar System, like the older Earth-centered one, was fated to be overthrown. The next revolution in astronomy began when a church organist decided to become a better musician.

A MUSICIAN LOOKS SKYWARD

Friedrich Wilhelm Herschel was born in 1738 in Hanover, Germany. His father, a musician in a military band, made sure that his four sons and two daughters were trained in music. He also inspired them with his interest in science.

Young Wilhelm, as he was known, was a good oboe player and

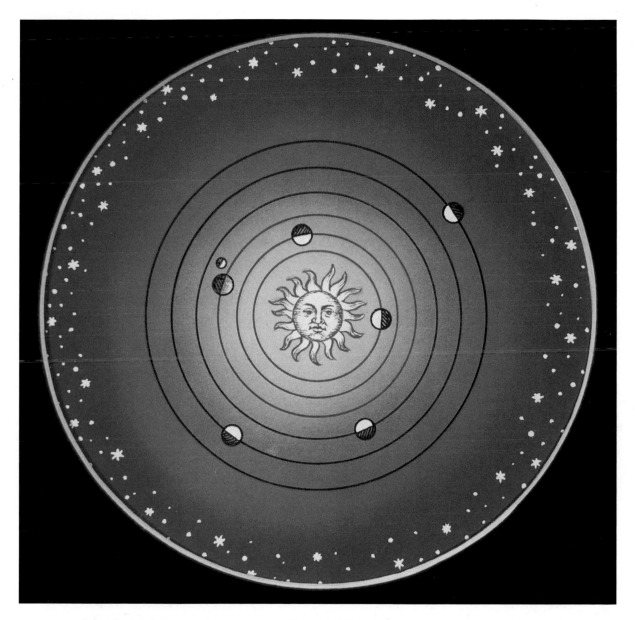

Copernicus's heliocentric, or Sun-centered, view of the Solar System included Mercury, Venus, Earth, Mars, Jupiter, and Saturn. Even the visionary Copernicus did not suspect the existence of other planets.

joined his father's band at the age of fourteen. Then, in the mid–1750s, the nations of Europe went to war. Wilhelm Herschel was nineteen when his band accompanied troops on a difficult and dangerous military campaign. Wilhelm soon decided that military life was not for him. His father got him discharged from military service, and he moved to England, where he supported himself by giving music lessons. He also changed his first name to William.

In 1766 Herschel settled in Bath, England, a popular and fashionable resort town. He got a job playing the organ and directing the choir in a church, although he continued to give lessons. Several of his brothers came from Germany to live with him for long periods. So did his sister Caroline, who managed his household and studied music.

Herschel liked to do things thoroughly. He believed that knowledge of mathematics would improve his musical skills, because musical harmony is based on the mathematical relationships among notes. In spite of his busy schedule, Herschel plunged into mathematics, studying at night. Soon he moved on to optics, a branch of mathematics and physics that deals with the properties of light. Optics is the basis of telescope construction, and Herschel grew interested in telescopes and astronomy. He began considering a question that puzzled astronomers: How far were the stars from Earth? Herschel was convinced that a detailed survey of the stars through a powerful telescope might help answer that question. An accurate map of the night sky would allow astronomers to notice and track any tiny movements that one star made relative to another. Such movements, if they occurred and could be measured, might give scientists a means of calculating the stars' distances. Ancient scientists such as the Greek astronomer and geographer Ptolemy had looked for such movements but had failed to find them. Herschel felt that the new telescopes would allow a more accurate survey of the heavens, and he decided to produce it.

His first necessity was a good telescope. Dissatisfied with the instruments he rented, Herschel began making his own telescopes in

Danish astronomer Tycho Brahe was so passionate about science that he fought a duel over a mathematical point, losing part of his nose to his enemy's sword (he fashioned a nose of wax and precious metals to replace it). Brahe's observations eventually proved Copernicus's heliocentric theory.

1773. He concentrated on the type of telescope known as a reflector because it uses mirrors to reflect light from the sky into an eyepiece. To achieve the results he wanted, Herschel knew that he would have to make large, extremely smooth, perfectly polished mirrors that had exactly the right curved shapes.

In the late eighteenth century, mirrors were not always made of glass. Some were made of a material called speculum metal, an alloy or blend of metals with a shiny surface. With dogged determination, Herschel began experimenting with different alloys, eventually settling on a mix of 71 percent copper and 29 percent tin. He also honed his skills as a polisher, sometimes working on a mirror for sixteen hours at a time. After many failed attempts, he produced an acceptable mirror five inches across and built a telescope. With it he made

William Herschel discovering Uranus— an event that not only excited scientists but inspired a poet. In his 1816 sonnet, On First Looking Into Chapman's Homer, *English poet John Keats wrote of the soul-stirring excitement felt by "some watcher of the skies/When a new planet swims into his ken [view]." He was referring to Herschel.*

several important discoveries, such as spotting two new moons of Saturn to be added to those that other astronomers had found since the invention of the telescope.

Soon Herschel began his grand project of surveying the sky. He also labored continually to build larger and better telescopes, turning his home into a mirror and telescope factory. His sister Caroline wrote, "[I]t was to my sorrow that I saw every room turned into a workshop." Herschel used the kitchen as a smelter, a furnace for melting metal. His activities were not just messy but occasionally dangerous. Once a large mirror mold cracked when Herschel and some workmen poured molten alloy into it. The hot metal spilled onto the stone floor, causing it to crack violently. Bits of rock flew through the air like bullets while hot metal pooled on the floor. Fortunately, everyone escaped injury.

Such accidents did not discourage Herschel. By 1778 he had made an excellent telescope with a mirror more than six inches across. He spent every night peering into the eyepiece, except when clouds or moonlight got in the way of his stargazing. Mealtime lost all meaning for him when he was at his telescope—Caroline simply fed him bites of food while he worked. She later wrote, "If it had not been sometimes for the intervention of a cloudy or moon-light night, I know not when my Brother (or I either) should have got any sleep."

On the night of March 13, 1781, Herschel was working at his never-ending project of listing the stars and their positions in the sky in order to compare them with earlier observations. He spotted an unusual new object in the constellation Taurus. As seen through his telescope, it was a disk, not a mere point of light like a star. He thought it was probably a comet, one of the balls of ice and gas that occasionally move through the Solar System trailing long, glowing tails. Herschel tracked the object's motion for several months, but it did not act like a comet. It appeared to be traveling around the Sun in an orbit like that of a planet. But of course it could not be a planet—everyone knew that Saturn was the outermost planet!

Before long, though, astronomers began to suspect that they did not know the Solar System as well as they had thought. Within a few months of Herschel's announcement of his discovery, investigators in Great Britain, Russia, and France had used astronomical observations and mathematical calculations to prove that the new object was indeed a distant planet. Herschel had done something that no one else in history had done—he had added a new planet to the Solar System. Later in 1781 the Royal Society, Britain's leading scientific organization, gave Herschel a medal and made him a member.

George's Star?

Everyone wondered what to call Herschel's new planet. The Royal Society asked Herschel to name it so that the French—rivals of the English in scholarship as well as politics—would not give it a name of their choosing. The French *did* suggest a name for the new planet. They wanted to call it Herschel. A few people used this name for a while, and it appears in at least one astronomy book from around 1800, but it did not stick. Herschel offered to name the planet *Georgium Sidus*, or George's Star, after King George III of Great Britain. Everyone outside Britain instantly rejected this idea, and people in different countries began calling the new planet by a variety of names. In England it was known for some years as the Georgian Planet. Eventually the name favored in Germany, Uranus, became established. Like the names of the other planets, Uranus comes from ancient Greek and Roman mythology. The sky god Uranus was the father of Saturn and one of the most ancient gods.

A NEW VIEW OF THE HEAVENS

The discovery of Uranus made Herschel famous. He gave up music and became a full-time astronomer with an income from the king. Although Herschel appreciated the king's generosity, the income was actually less than he had earned as a working musician, so he added

Telescopes made by William Herschel, the discoverer of Uranus, were greatly prized. One of the finest was this 25-foot instrument with a 24-inch mirror, made for the king of Spain. The frame rotated on its base, and a system of pulleys and cables let the operator aim the telescope at any point in the sky.

to it by manufacturing and selling telescopes—the king of Spain, for example, paid Herschel a small fortune for a 25-foot-long telescope built for the Madrid Observatory.

For his own use Herschel built a 40-foot-long telescope with a 48-inch mirror, which remained the largest telescope in the world for more than fifty years. By his death in 1822, Herschel had completed four surveys of the stars, discovered the existence of binary or double stars (pairs of stars that revolve around each other), determined that the stars are

A STAR-STRUCK FAMILY

William was not the only member of the Herschel family to become a well-known astronomer. Two other Herschels also studied the heavens.

Caroline Herschel, William's sister, helped her brother by keeping records of his observations and by performing complex mathematical calculations connected with his work. But she did more than aid the discoverer of Uranus. She also used a telescope to make her own survey of the night sky. In the course of this project, Caroline Herschel discovered several nebulae and a number of comets. After William's death, she returned to Germany, where she continued to work on a detailed catalogue of her brother's discoveries and her own. Before her death in 1848, she won prizes from scientific societies and a gold medal for achievements in science from the king of Prussia.

John Frederick William Herschel was William Herschel's son. He was born in 1792, eleven years after farther away than scientists had thought, discovered infrared light, and developed new ideas about the shape of the galaxy and the Universe.

William Herschel was chiefly remembered, however, as the discoverer of Uranus. By adding a seventh planet to the Solar System, he had made scientific history and revolutionized the way people thought about outer space. Herschel the backyard skywatcher had shown that the Solar System was bigger, more complex, and more surprising than the leading scientists of his time had ever dreamed.

the discovery of Uranus. By that time, John Herschel's father had become one of the most famous astronomers of the day, and the boy grew up in a scientific atmosphere. At first he was interested in mathematics, and as a young man he published several papers on the subject. In 1816 he took up the family passion, astronomy. With his father's help he built a powerful telescope that allowed him to make detailed observations. Realizing that the skies of the southern hemisphere were practically unexplored compared with those visible to Europeans, John Herschel took his telescope to South Africa, where he spent four years making a celestial survey. He also studied the movements of double stars. During his long and successful scientific career, John Herschel made advances in chemistry and photography, but like his father he is most honored for his contributions to astronomy.

2

Voyager 2 Visits the Outer Solar System

O nce Uranus was added to the Solar System, astronomers around the world turned their attention and their telescopes toward the distant planet. They discovered a few important things, but most of what scientists now know about Uranus has been learned since 1986. In that year a spacecraft named *Voyager 2* passed briefly near Uranus. It is the only spacecraft from Earth that has ever visited the seventh planet.

EARLY STUDIES OF URANUS

Astronomer William Herschel naturally took an interest in the planet he had discovered. In 1787 he discovered two moons orbiting Uranus. They did not receive names until around 1852, when Herschel's son named them Titania and Oberon, after characters in one of William Shakespeare's plays. Two moons discovered the previous year by British astronomer William Lassell also received literary names: Ariel from a Shakespeare play and Umbriel from a poem by British writer Alexander Pope. American astronomer Gerard Kuiper followed the Shakespearean naming tradition when he discovered a fifth moon, Miranda, in 1948.

The Voyager 2 *spacecraft approaches Uranus's moon Miranda in this computer-generated image. The amber-colored bar is the probe's extendable boom, a movable arm carrying scientific equipment.*

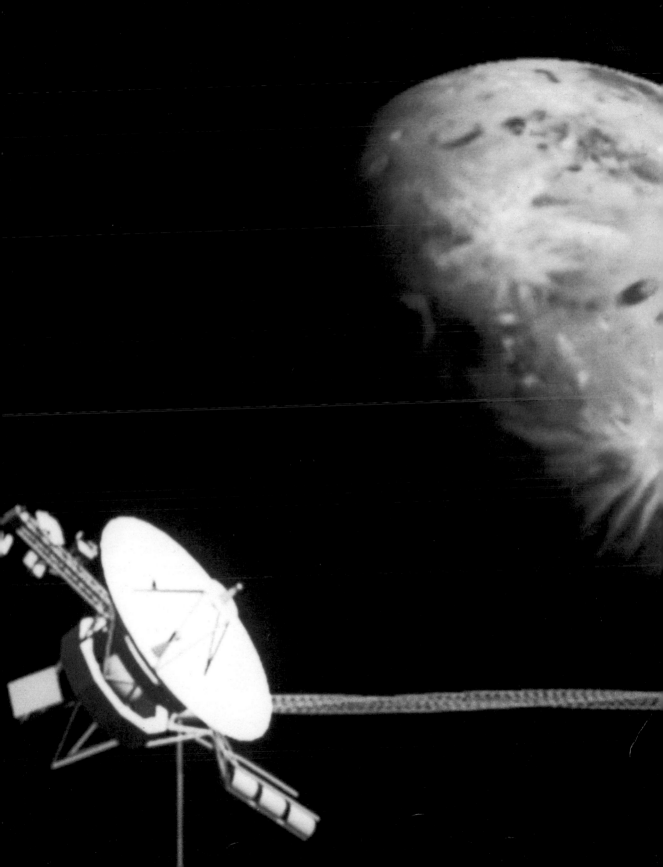

Uranus is more than nineteen times farther from the Sun than Earth. Its location in the outer part of the Solar System makes it difficult for astronomers on Earth to study. In fact, reviews of old astronomical observations show that both Tycho Brahe and Galileo had spotted Uranus before Herschel, but it was so small and faint that neither was able to identify it as a planet. As telescopes improved, however, astronomers learned more about the seventh planet. By the twentieth century scientists had determined that Uranus is the third-largest planet in the Solar System, after Jupiter and Saturn, but they had not yet managed to measure its exact size. Then, in 1977, they got a chance to do so. Uranus was due to pass in front of a star, an event known as an occultation. Scientists already knew the rate at which Uranus travels in its orbit. If they could measure how long Uranus blocked the star, they could calculate the planet's diameter.

Several teams of scientists observed the occultation from different points on Earth's surface—this increased the chances that at least one group would have clear skies. One team even recorded the occultation from a high-flying research airplane. The experiment gave a diameter

Can You See Uranus?

Uranus is sometimes visible in the night sky, depending upon the positions of Uranus and Earth in their orbits. If you know exactly where to look and have very good eyesight, you can spot the seventh planet, which looks like a very faint star to the naked eye. Through a good pair of binoculars, Uranus appears larger and brighter, with a bluish tint. If you look at Uranus through a telescope, you will see it as a disk rather than a point of light.

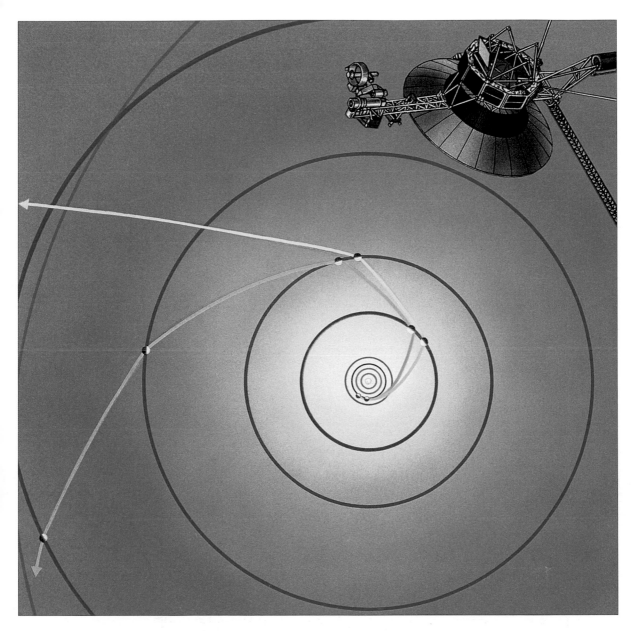

Launched in September 1977, Voyager 1 *(yellow track) visited Jupiter and Saturn before heading out of the Solar System.* Voyager 2, *launched several weeks earlier (orange track), continued on to Uranus and Neptune—thanks to the efforts of determined space scientists.*

for Uranus of almost 32,000 miles (more than 51,500 kilometers), a little more than four times Earth's diameter. Surprisingly, the star's light had appeared to flicker several times before the planet passed in front of it and again afterward. Most experts believed that this meant that Uranus had rings of dust or small particles around it, like the famous rings of Saturn. The star's light had "flickered" when the rings briefly blocked it. By the mid–1980s, observations of other occultations indicated at least nine rings around Uranus. But an unfolding adventure millions of miles from Earth was about to give scientific knowledge of the seventh planet a great boost.

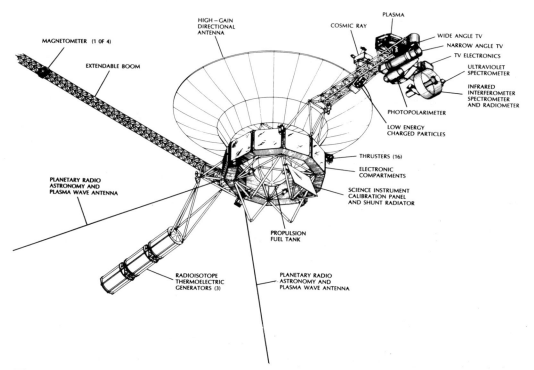

The 1,820-pound (839-kilogram) Voyager *spacecraft carried 231 pounds (105 kilograms) of scientific and communications gear. The probes gathered information not just about planets and their moons but also about interplanetary space.*

THE *VOYAGER* MISSIONS

In the mid–1960s, before astronauts had even reached the Moon, American scientists were planning something they called the Grand Tour. It would be a voyage through distant reaches of the solar system, but no astronauts would make the trip. Instead, probes loaded with sensors would gather information and radio it back to Earth.

Astronomers had figured out an efficient way for spacecraft to reach the giant planets of the outer Solar System. Once a craft reached Jupiter, it could swing around the massive planet, picking up speed from the force of Jupiter's gravity. This maneuver, called gravitational assist or the slingshot effect, would send the craft hurtling toward its next destination. The U.S. National Aeronautics and Space Administration (NASA) wanted to send a nuclear-powered probe to Jupiter and then use the slingshot effect to send it on to Saturn, Uranus, and Neptune, the eighth planet, which astronomers had discovered in 1846. If NASA could launch the probe in 1977, the four planets would be lined up perfectly for a probe to visit them all in a twelve-year mission.

When NASA first began planning the Grand Tour, no probe had gone farther than Mars, but scientists felt sure they could meet the technological challenges of making a probe that would function much farther from Earth for years on end. The financial and political challenges, however, proved too tough for them.

Although the United States had reached the Moon, it had no national plan for continuing space exploration. Congress refused to spend a billion dollars on the Grand Tour. It did give NASA $360 million for a less ambitious mission: sending two probes to Jupiter and Saturn.

Those probes, *Voyager 1* and *Voyager 2*, lifted off in 1977. By that time the Grand Tour was on again—sort of. One probe was supposed to visit Io, a moon of Jupiter, and Titan, a moon of Saturn. Its flight path would not let it make the Grand Tour. But NASA scientists had

An artist's view of Voyager 2 *nearing its closest approach to Uranus. Back on Earth, scientists eagerly awaited their first—and, so far, their only—close glimpse of the seventh planet.*

figured out that the other probe could use the gravitational assist of Saturn to complete the Grand Tour at very little extra cost. In 1981 the project managers adjusted *Voyager 2*'s flight path to slingshot it around Saturn and head it toward Uranus. NASA hoped that the probe and its equipment would survive a much longer mission than the one for which they had been designed.

A Voyager probe moves into interstellar space in this artist's conception. If Voyager 2 *stays on its present path, it will pass close to a star called Ross 248 in a little more than 40,000 years.*

SWINGING PAST URANUS

For four and a half years *Voyager 2* traveled from Saturn toward Uranus. During that time, scientists struggled to fix problems with the probe's radio equipment, computer memory, and sensor platform. Although the probe was millions of miles away and could communicate with its Earth-bound controllers only through faint radio signals, the scientists succeeded brilliantly. They reprogrammed the probe's computers and learned how to use short rocket bursts to keep the spacecraft stable while its cameras were taking pictures. By the time *Voyager 2* approached Uranus, it was working better than ever.

In January 1986, 205 years after Herschel discovered Uranus, the world got its first close look at the seventh planet. Uranus was then 1.75 billion miles (2.8 billion kilometers) away from Earth—so far that *Voyager 2*'s radio signals took nearly three hours to reach Earth. At its closest approach to Uranus, the spacecraft was 50,700 miles (81,600 kilometers) above the top of the planet's thick cloud cover. *Voyager*'s cameras took thousands of photographs. Other instruments gathered information about such features as temperature and chemical makeup.

Scientists back on Earth had been waiting years for the encounter with Uranus. They desperately wanted to make the most of their first chance to study the planet and its satellites. To their delight, *Voyager 2* kept working perfectly and even managed to capture images of the moons Titania, Oberon, Ariel, Umbriel, and Miranda as it swung around Uranus. Then, flung onward by a boost from Uranus's gravity, the far-traveling spacecraft headed out toward Neptune. Long before it reached the final planned encounter on its Grand Tour, scientists were hard at work analyzing the information it had sent them about Uranus. The data *Voyager 2* had gathered would keep them busy for years, unraveling the secrets of the seventh planet.

ON TO THE STARS

Voyager 2 did reach Neptune in 1989, and it sent home pictures and data, just as it had for Uranus. Where did the spacecraft go next?

Both *Voyager 1* and *Voyager 2* are following flight paths that will eventually carry them out of the Solar System. Their nuclear-powered engines will someday give out. Their instruments and their radio links with Earth will fail. Blind, deaf, and cold, the *Voyager* spacecraft will drift on toward the edge of the Solar System and beyond it into interstellar space. Someday, perhaps, the *Voyager* spacecraft will be studied by beings who dwell on other planets. If this ever happens, those beings may play a recording that was placed in each spacecraft. Made of copper and gold and designed to last for a billion years, this recording carries pictures of Earth and its inhabitants and messages in dozens of languages. President Jimmy Carter of the United States sent this message: "This is a present from a small distant world. . . . We are attempting to survive our time so we may live into yours. We hope someday, having solved the problems we face, to join a community of galactic civilizations. This record represents our hope and our determination, and our good will in a vast and awesome universe."

3

UNDERSTANDING URANUS

The *Voyager* 2 flyby of Uranus answered many questions about the seventh planet. At the same time, it surprised scientists by revealing that some of their ideas about the planet were wrong.

REVOLUTION AND ROTATION

The spacecraft's instruments confirmed earlier measurements of Uranus's distance from the Sun. The mean distance, the midpoint between the planet's closest and farthest distances from the Sun during its orbit, is 1,782 million miles (2,868 million kilometers).

All planets have two main kinds of motion: revolution and rotation. These determine the length of its years and days. A planet's period of revolution is the time it takes to make one circuit around the Sun—in other words, a year. Uranus travels once around the Sun every eighty-four Earth years, which means that a Uranian year is eighty-four Earth years long.

Uranus's year may last a human lifetime, but its day is shorter than a day on Earth. A planet's day is its period of rotation, the amount of time it takes for the planet to spin completely around.

An infrared image of Uranus taken by the Hubble Space Telescope shows the planet's rings as well as several clouds (orange spots) the size of earthly continents. The gray area is Uranus's south pole.

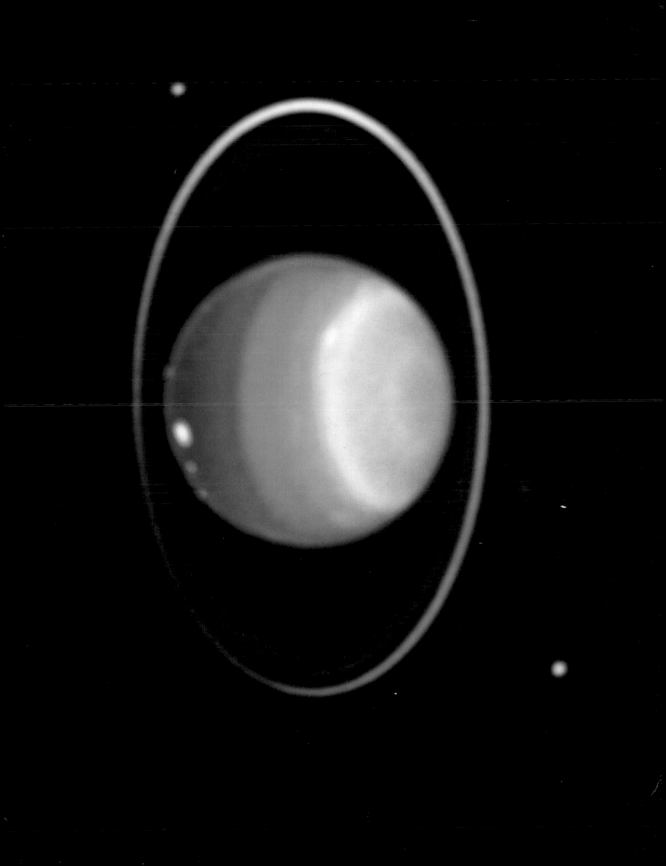

The Sun (left) and planets, shown nearly to scale. From left to right the planets are Mercury, Venus, Earth, Mars, Jupiter, Saturn, Uranus, Neptune, and Pluto. Uranus is one of the four "gas giants," all much larger than Earth.

Before *Voyager* 2, scientists had little luck measuring Uranus's day. The usual method of measuring a planet's period of rotation is to focus on a feature on the planet's surface and see how long it takes for that feature to come around to exactly the same point again. Uranus is so far from Earth that astronomers had trouble focusing on its surface. And when they managed to do so, the planet's dense cover of clouds presented a blank, featureless face. *Voyager* 2, however, gave scientists another way to determine the length of a Uranian day. The probe measured variations in the strength of Uranus's magnetic field, which rotates along with the planet. Sensors aboard the spacecraft recorded the fact that the magnetic field rotates once every 17 hours, 14 minutes. This means that a Uranian day is almost three-fourths as long as an Earth day.

Another puzzle connected with Uranus's rotation remains unsolved even after *Voyager* 2. A planet rotates on its axis, an imaginary line that runs between its north pole and its south pole. Picture the axis of a planet as a pencil stuck through the center of an orange. Earth's axis is nearly upright, tilted just a little. If the orange were Earth, it would be spinning around on the point of the pencil while also revolving around the Sun. In most other planets, the axis is also upright, or only slightly tilted. But Uranus's axis is tilted so far that the planet rotates on its side. If Uranus were an orange with a pencil for an axis, the orange would roll along with the ends of the pencil sticking out of each side like handles. No other planet rotates at this angle, and scientists do not know why Uranus does it. Some think that when the Solar System formed, all of the planets had fairly vertical axes. At some point Uranus was struck by something large, perhaps a comet or even another planet, and the collision knocked Uranus over onto its side.

A GAS GIANT

Jupiter, Saturn, Uranus, and Neptune are sometimes called the "gas giants." They are enclosed in very deep atmospheres, or layers of gas. In contrast, Earth and the other planets are rocky worlds with much shallower atmospheric layers. In addition to having thick atmospheres, the gas giants are much larger than the other planets of the Solar System. Although Uranus is not nearly as large as Jupiter or Saturn, it is big enough to contain sixty Earths.

Before *Voyager 2*, scientists thought that Uranus was made up mostly of hydrogen and helium. The spacecraft's survey proved them right, except that there was more hydrogen and less helium than they expected—83 percent of the planet's atmosphere is hydrogen, while about 15 percent is helium. About 2 percent of the atmosphere is methane gas. When light from the distant Sun strikes Uranus, the methane absorbs the red portion of the sunlight. The fact that no red light is reflected back from the surface is what gives Uranus its distinctive greenish blue color. *Voyager 2*'s sensors also detected small amounts of water, acetylene, ammonia, ethane, and a few other chemicals in the Uranian atmosphere.

Scientists who had studied the *Voyager* information on Jupiter and Saturn expected Uranus to be a lot like those planets. In particular, they thought its structure would resemble theirs. Jupiter and Saturn appear to have cores, or centers, of heated rock. Each core is

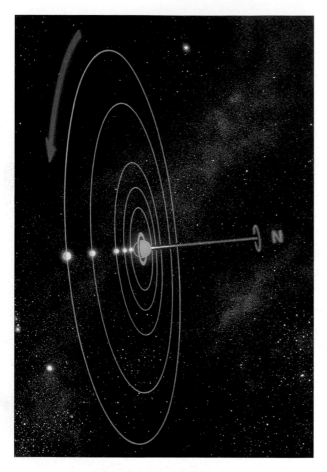

The red line on the left shows the direction of Uranus's movement in its orbit. The planet is tilted at an angle of 98 degrees to the plane of its orbit, which means that it lies on its side, with its moons revolving around its Equator.

surrounded by a thick ocean of liquid hydrogen. The third and outermost layer of each planet is an atmosphere made up of various gases. Experts who examined the *Voyager 2* data on Uranus thought at first that the seventh planet also had a three-part structure of rocky core, liquid ocean, and gaseous atmosphere. Uranus's liquid layer, they believed, contained large amounts of water along with the hydrogen.

Further study, however, led to a new idea about the structure of Uranus. The planet spins on its axis very rapidly for its size. Scientists realized that if it had a liquid layer, the rapid spinning would cause the ocean to bulge outward around the planet's Equator, or center. But *Voyager 2* measured very little bulge at the Equator of Uranus. For this reason, many scientists now think that Uranus may have only two layers. At its heart is a rocky core perhaps not

Dayglow

Voyager 2's instruments detected an unusual light shining out of the side of Uranus that faced the Sun. This light is invisible to the human eye because its wavelengths are in the ultraviolet range. Scientists think the light is caused by hydrogen atoms high in Uranus's atmosphere giving off energy in the form of ultraviolet light. They call this radiation dayglow because it only occurs on the daylight side of the planet.

much larger than Earth. Around the core is a very thick and heavy atmosphere. The upper part of the atmosphere is a fine haze or smog. It consists of extremely tiny particles of chemicals called hydrocarbons, created when sunlight causes chemical reactions in the atmosphere. Below the haze is a layer of clouds made of tiny frozen methane crystals floating in the atmosphere. A few researchers have suggested that Uranus may not even have a rocky core. According to their theories, the planet is a very dense ball of mixed gases with no internal layers.

CONDITIONS ON URANUS

Like the other gas giants, Uranus is so unsuited to human life that it is hard to imagine people visiting the planet. In one way, though, Uranus is not too different from Earth. Its gravity is 91 percent of Earth's. If you weighed 100 pounds on Earth, you would weigh 91 pounds on Uranus. You could jump a little higher there than at home —if, that is, you had a place to stand and somewhere to jump to. But even if you tried to fight your way through the poisonous atmosphere

Uranus Fact Sheet

Mean Distance from Sun: 1,782 million miles
  (2,868 million kilometers)
Period of Revolution: 84 Earth years
Period of Rotation: 17 hours, 14 minutes
Diameter: 31,700 miles (51,500 kilometers)
Atmosphere: 83% hydrogen, 15% helium, 2% methane
Average Temperature: −366 degrees Fahrenheit
  (−212 degrees Celsius)
Surface Gravity: 0.91% Earth gravity

toward the planet's core, the weight of the gas above you would bear down with crushing pressure.

Because of the way its axis is tilted, Uranus always points one or the other of its poles or its Equator toward the Sun as it revolves in its orbit. For one-fourth of each Uranian year (twenty-one Earth years), the planet's south pole receives nonstop sunlight. For the next one-fourth of the year, the Equator is positioned to receive direct sunlight. For another one-fourth of the year, the south pole is completely dark while the north pole basks in sunlight. "Basks" may not be the right word, however. Sunlight on Uranus is not at all like the light that warms and illuminates Earth. Seen from Uranus, the Sun is not a fiery disk—it is merely the brightest and largest star in the sky. The average temperature close to the planet, in its lower atmosphere is around −366 degrees Fahrenheit (−212 degrees Celsius), much colder than anything ever recorded on Earth.

On Jupiter and Saturn, terrific winds lash the atmosphere, sending huge belts of cloud streaming around the planets. Buried in those belts are vast circular storms. Some of these storms are many times larger than Earth and are known to have lasted for centuries. *Voyager 2's*

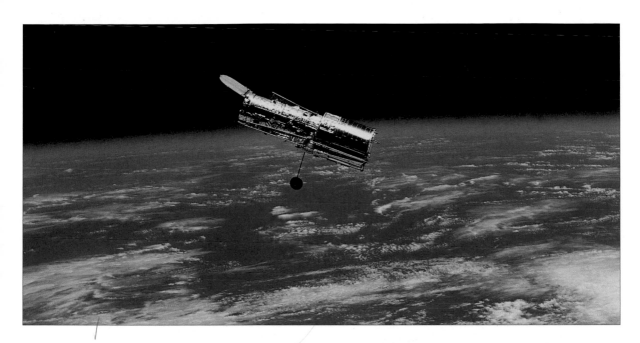

High in orbit above the Indian Ocean, the Hubble Space Telescope separates from the space shuttle Discovery *in 1997. Hubble gave Earth-based astronomers a powerful new tool for studying distant planets such as Uranus.*

images of Uranus revealed little evidence of storms. In 1999, however, NASA released the results of a study of Uranus conducted through the Hubble Space Telescope (HST), which orbits Earth. Astronomers using the HST were able to observe and film dramatic changes in Uranus's atmosphere as the planet's northern hemisphere began to move out of its long winter toward its equally long summer. Waves of huge storms swept across Uranus's surface.

Like the other gas giants, Uranus is a windy world. Winds in the upper atmosphere circle the planet at speeds of between 90 and 360 miles (145 and 576 kilometers) per hour. The lower end of this range is not too different from speeds recorded for Earth's jet streams, the high-altitude winds that create much of our weather. They typically blow at around 110 miles (177 kilometers) per hour. Uranus's faster winds, though, are like everything else on the planet: unearthly.

4

MOONS AND RINGS

Before *Voyager 2* visited Uranus, the planet was known to have five moons and nine encircling rings. Together the planet, moons, and rings form what astronomers call the Uranian system, its parts held together by the pull of gravitational forces. *Voyager 2* gathered a lot of information about Uranus's moons and rings—including the fact that there are more of them than scientists had thought.

MAJOR MOONS

The five largest Uranian satellites are big enough to have been discovered from Earth through telescopes before the *Voyager 2* mission. Still, all five are quite a bit smaller than Earth's Moon, which measures 2,160 miles (3,476 kilometers) in diameter. Scientists expected Uranus's moons to be rather dull, with no signs of geological activity on their surfaces. Moons so small, they thought, were merely chunks of rock and ice that had never had enough radioactive material to produce internal heat. Molten rock had never flowed across their surfaces, volcanoes had never erupted, ice had never melted and caused floods. All that had ever happened on these dead little worlds was

Pieced together from many Voyager 2 *images, this view of the Uranian system shows how Uranus would have looked to a passenger aboard the probe, with five of the planet's moons in the foreground.*

Parts of Oberon's surface resemble the cratered highlands of Earth's Moon, pocked by impacts during the formation of the Solar System more than 4 billion years ago.

that meteors, rocks, and dust had crashed into them for 4.6 billion years, since the birth of the Solar System. Their surfaces would be pocked with craters large and small from these impacts.

Oberon, the second largest of the five known moons, was about what scientists expected. Its surface was covered with craters. One mountain towered 4 miles (6 kilometers) or more above the surrounding terrain. It was probably created when a huge meteor smashed into Oberon. The impact sent a fountain of rock skyward. As scientists studied photographs of Oberon, however, they noticed something unexpected. The floors of some craters were smooth and dark. It looked as if some liquid matter had flowed upward from within the moon and then hardened, but too recently to have been broken up by meteor impacts. This suggested that perhaps the moons of Uranus were more complex than anyone had thought.

Titania offered still more surprises. Vast, deep cracks crisscross its surface, signs of powerful earthquake-like movements in its past. Internal geological forces, not meteor impacts, caused these movements. A hot, molten core might have moved sections of Titania's hard outer crust, breaking it into fault valleys. Another possibility is that radioactive heat from the moon's core once melted massive amounts of ice, which filled narrow crevices and froze. Expanding as it froze, the ice split the moon's surface into huge, zigzagging cracks. The biggest valleys on Titania are three times as deep as the Grand

The level plains of Umbriel stretch toward the horizon in this artist's visualization of that moon's surface. Ringed Uranus looms in the sky.

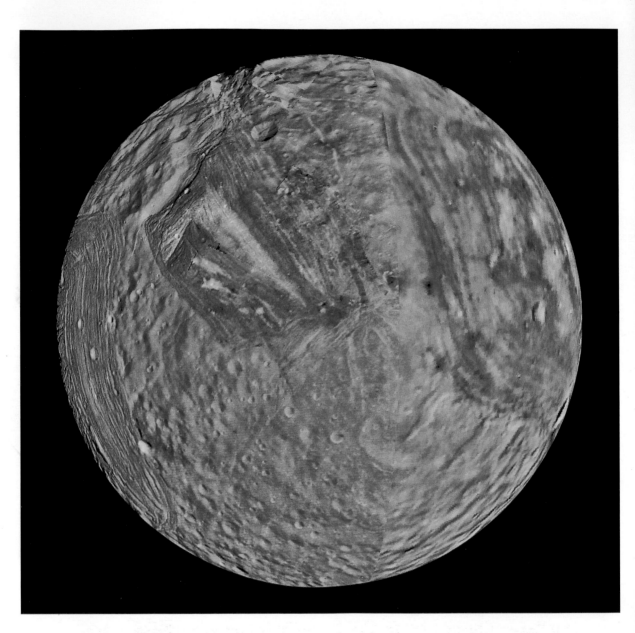

Miranda, the smallest of Uranus's major moons, has the most dramatic geology. The dark "bite" out of its upper rim is the silhouette of an immense valley.

Canyon and more than 930 miles (1,500 kilometers) long. On Earth such a chasm would stretch halfway across the United States.

Umbriel shows little sign of geological activity. Its most unusual feature is its very dark surface, broken only by a few patches of bright ice. Scientists do not know why Umbriel is much darker than the other Uranian moons, but one possibility is that a cloud of dark dust somehow formed nearby, perhaps through the collision of two small moons, and coated Umbriel.

Ariel is the lightest and brightest of the Uranian moons. Some experts think that this is because its surface is the youngest. Although Ariel has many impact craters, most are fairly small. Scientists think that enormous amounts of slushy ice have flowed across Ariel's surface, covering most of the large craters formed during the intense meteor bombardment of the young Solar System. Freezing shattered the surface, creating an even more massive system of faults than Titania's. Some of the valleys on Ariel are 20 miles (30 kilometers) deep.

Miranda, the smallest of the five major moons, was the most astonishing. *Voyager 2* passed between Uranus and Miranda, traveling within 17,400 miles (28,000 kilometers) of the moon's surface—closer than it came to any other planet or moon in its long journey. The spacecraft captured images that made scientists' jaws drop. Little Miranda's surface was as varied and unusual as that of any world in the Solar System. Craters large and small dot the moon. Huge canyons 12 miles (19 kilometers) deep encircle it. Three large, bright, oval-shaped regions are carved into rows of parallel grooves, ridges, and cliffs. One of them has a large, bright, V-shaped feature nicknamed the Chevron.

Researchers do not know what forces twisted Miranda's surface into these distinctive forms. One theory is that the moon broke apart after a violent impact, and as gravity drew the pieces back together, chunks of Miranda's inside and outside were jumbled together.

Another idea is that melted ice from Miranda's interior began to flow upward, creating the bright oval regions, but the moon cooled before the ice could cover its entire surface. Many experts think that the heat that once powered Miranda's geological upheavals came from friction inside the moon caused by the competing gravitational pulls of Uranus and Ariel, Miranda's neighbors. Whatever its cause, Miranda packs more variety into a small area than anything else in the Solar System. Upon seeing the first images of the moon, one scientist said, "It looks like a satellite designed by a committee."

MINOR MOONS

Voyager 2 revealed ten new moons. All are small, fairly dark, and closer to the planet than the big five moons. The spacecraft's flight path allowed it to obtain one picture of Puck, the largest of the new moons, but it got no close views of the others. Scientists believe that, like Puck, they are covered with craters and charcoal-colored material.

That dark material is a topic of debate among space scientists. Some believe that it is methane ice that was originally light in color. Over millions of years, the energy of Uranus's magnetic field has caused chemical reactions to coat the methane ice with a black hydrocarbon compound that has been called "star tar." Others think that the dark material is a carbon compound that remains unchanged from the formation of the Solar System. Either way, the blackish color leads researchers to think that the material is some kind of carbon, an element scientists believe is necessary for life. Its presence on the Uranian moons suggests that there is plenty of carbon in the Solar System.

Voyager 2 brought the total number of satellites around Uranus to fifteen. During the late 1990s, scientists located six more satellites. With a total of twenty-one, Uranus had more moons than any other planet in the Solar System—until 2000, when astronomers announced that they had found four new moons around Saturn, bringing that planet's total to twenty-two. Of the new moons found around Uranus, five are very small and very far from the planet. Like the earlier-known moons, all the recent discoveries have been named for characters from the works of Shakespeare and Pope.

An earthquake—or moonquake—on Miranda threatens a human colony in this artist's vision.

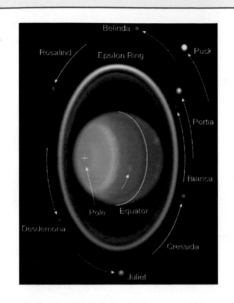

The paths of some of the Uranian moons are illustrated in this image of the planet and its rings. The Epsilon ring is the largest in the Uranian ring system.

THE MOONS OF URANUS

Here are Uranus's twenty-one moons, starting closest to the planet. (Distances are rounded to the nearest thousand miles or kilometers.)

NAME	DISTANCE FROM URANUS	DIAMETER	YEAR DISCOVERED
Cordelia	31,000 mi (50,000 km)	8 mi (13 km)	1986
Ophelia	33,000 mi (53,000 km)	10 mi (16 km)	1986
Bianca	37,000 mi (59,000 km)	14 mi (22 km)	1986
Cressida	38,000 mi (61,000 km)	20 mi (32 km)	1986
Desdemona	39,000 mi (63,000 km)	18 mi (29 km)	1986
Juliet	40,000 mi (64,000 km)	26 mi (42 km)	1986
Portia	41,000 mi (66,000 km)	34 mi (55 km)	1986

Name	Distance from Uranus	Diameter	Year Discovered
Rosalind	43,000 mi (69,000 km)	17 mi (27 km)	1986
Belinda	46,500 mi (75,000 km)	21 mi (34 km)	1986
1986U10*	47,000 mi (76,000 km)	25 mi (40 km)	1999
Puck	53,000 mi (86,000 km)	48 mi (77 km)	1985
Miranda	80,000 mi (130,000 km)	146 mi (235 km)	1948
Ariel	118,000 mi (190,000 km)	359 mi (578 km)	1851
Umbriel	165,000 mi (266,000 km)	363 mi (584 km)	1851
Titania	270,000 mi (435,000 km)	489 mi (787 km)	1787
Oberon	361,000 mi (581,000 km)	472 mi (760 km)	1787
Caliban	4,448,000 mi (7,168,000 km)	25 mi (40 km)	1997
Stephano	4,928,000 mi (7,951,000 km)	9 mi (14 km)	1999
Sycorax	7,572,000 mi (12,186,000 km)	50 mi (80 km)	1997
Prospero	10,272,000 mi (16,531,000 km)	12.5 mi (20 km)	1999
Setebos	10,962,000 mi (17,641,000 km)	12.5 mi (20 km)	1999

*Not yet named.

A computer-generated image based on **Voyager** data shows four sections of the Epsilon ring. The computer added light color to the ring material, which in reality is quite dark.

Uranus's ring system is smaller and fainter than Saturn's better-known rings. Discovering some of the dim Uranian rings from Earth was a triumph of astronomy.

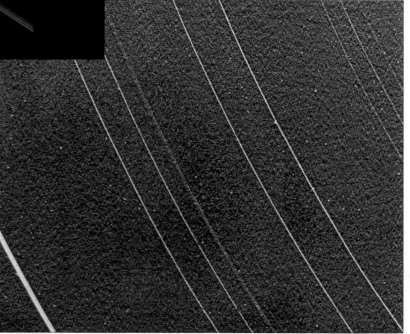

THE RINGS OF URANUS

Voyager 2 photographed and measured the nine known rings of Uranus. It also found at least two new ones. Researchers are still studying images from the spacecraft that hint at the existence of additional faint, narrow rings.

The largest ring is also the farthest from Uranus. Called the Epsilon ring, it is 32,000 miles (51,000 kilometers) from the planet. Its width varies from 14 to 58 miles (22 to 93 kilometers). The Epsilon ring lies between the orbits of Cordelia and Ophelia, Uranus's small innermost moons. They are nicknamed the "shepherd moons" because their gravity keeps particles from leaving the ring. Like herders guiding a flock of sheep, they keep the ring together. Scientists expected to find shepherd moons for all the rings, but no others have yet been located. They may be too small and too close to the planet to be easily spotted.

The 1999 study that discovered storms on Uranus also discovered that the rings wobble slightly, "like an unbalanced wagon wheel," as NASA scientists described the motion. The cause of the wobble is unknown. The gravitational pull of the planet's many small moons may be enough to disturb the rings.

Uranus's rings consist of chunks of rock and ice. They contain much less dust—and less material overall—than the ring systems of Jupiter and Saturn. Although collisions between rocks in Uranus's rings constantly produce dust, the dust is sucked down into Uranus's thin upper atmosphere, which stretches all the way out into the ring system. Uranus's rings probably consist of material thrown into space when comets or asteroids struck its satellites. The small inner moons may also be fragments of such dramatic impacts. Like the inner moons, the ring material is dark. Scientists believe that it is covered with the same carbon compound that coats the small moons.

5

URANIAN MYSTERIES

We have learned much about the seventh planet since William Herschel spotted it through a telescope in his backyard. Still, planetary scientists have more questions about Uranus than answers. Does the planet have a rocky core? Why is it tilted on its side? One puzzle, concerning the orbit of Uranus and the structure of the Solar System, has teased astronomers since soon after Uranus was discovered. Another mystery involves Uranus's magnetic field and its relation to the planet.

A WAYWARD WORLD AND A TENTH PLANET?

William Herschel's discovery doubled the extent of the Solar System —Uranus is twice as far from the Sun as Saturn. This knowledge led to an even greater expansion of the Solar System.

A century earlier, mathematician Isaac Newton had described the laws of motion, force, and gravity that control all moving bodies. The system of laws and principles he outlined is known as Newtonian physics. It explained how the planets orbit the Sun, and it let scientists predict their exact courses. Astronomers eagerly applied Newtonian physics to the newly discovered Uranus and found, to their surprise,

Uranus's major moons (left to right, Umbriel, Miranda, Oberon, Titania, and Ariel) hang above the planet in this computer illustration.

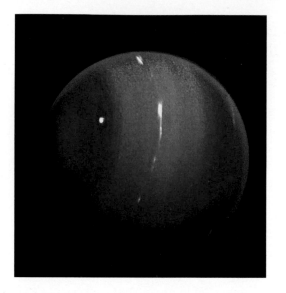

Neptune, the eighth planet, was discovered because it caused irregularities in Uranus's orbit. Neptune is a gas giant like Uranus, but the higher percentage of methane in its atmosphere makes it look bluer.

that the seventh planet was not following its predicted course. Nothing was wrong with Newtonian physics, so what was wrong with Uranus? By the 1840s scientists suspected that Uranus was affected by the gravitational pull of an undiscovered planet farther from the Sun. They used Newtonian physics to predict where such an eighth planet might be found, if it existed. In 1846 astronomers located the eighth planet and named it Neptune.

The mystery of Uranus's motion had been solved. Or had it? Scientists soon found that even after allowing for Neptune's presence, Uranus was *still* not behaving as it should. Many suspected that yet another planet lay beyond Neptune, disturbing the orbits of both Neptune and Uranus with its gravity. Not until 1930, however, did American astronomer Clyde Tombaugh succeed in locating small, hard-to-see Pluto, the ninth planet. The discovery of Uranus had led directly to the finding of two more worlds.

Pluto, however, turned out to be too small to have much effect on Uranus. As a result, some space scientists claimed that an unknown tenth world lurked undiscovered on the far edge of the Solar System. They dubbed it Planet X and launched a search for it. Today a few planet hunters continue that search, arguing that a tenth planet exists somewhere. Many scientists, however, now believe that the irregularities long thought to exist in the orbits of Uranus and Neptune may not exist at all—they may simply be errors in measuring or recording.

MYSTERIOUS MAGNETISM

Scientists working from Earth had not managed to find a magnetic field around Uranus. Such a field is created when a planet has heat at its core. The heat causes particles called electrons to split off of atoms. Electrons have an electric charge, and when the planet rotates, their movement creates an energy field around the planet called magnetism. This force makes compass needles point toward the poles of the magnetic field and attracts metal to magnetically charged objects.

Voyager 2 revealed that Uranus does have a magnetic field. It is more powerful than Earth's. On Earth and the other magnetically charged planets, the magnetic field is aligned fairly closely with the planet's axis, so that the north and south magnetic poles are not far from the north and south poles around which the planet rotates. The center of the magnetic field also passes close to the center of the planet. But Uranus's magnetic field does not line up at all with its axis of rotation. The planet's magnetic poles are closer to its equator than to its north and south poles. In addition, the axis between Uranus's magnetic poles is 4,800 miles (7,700 kilometers) from the planet's center. Scientists think that Uranus's cockeyed magnetic field may be produced near the planet's surface, not at its core like the magnetic fields of other planets. They do not yet know, however, what conditions give rise to the magnetic field.

A Problem for Astrologers?

William Herschel's 1781 discovery of Uranus surprised scientists who had not dreamed that the Solar System could contain unknown worlds. It also challenged followers of the completely unscientific practice called astrology.

Known in some cultures since ancient times, astrology is a form of fortune-telling based on the belief that the position of the planets at the moment a person is born somehow influences that person's character or fate, or that the position of the planets makes certain days favorable for particular activities. All of the various astrological systems placed great importance on the traits linked with the Sun, the Moon, and each of the planets—each of the *five* planets, that is, because only Mercury, Venus, Mars, Jupiter, and Saturn were known when astrology developed. (Earth did not count because ancient peoples did not think of it as a planet.)

So how did astrologers respond to Herschel's new planet? You might think that the sudden discovery of Uranus would overthrow a fortune-telling system that had not even managed to foretell the discovery! You might also expect that believers in astrology would become doubtful and wonder how the system could have worked for thousands of years without taking Uranus into account. But some people seek supernatural or mystical guidance in life, even if it goes against common sense and reason, and a market for astrological fortune-telling

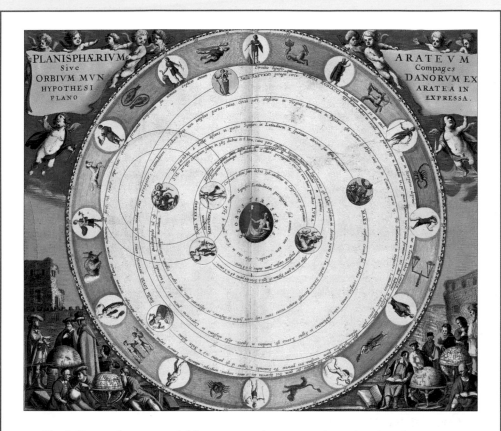

Earth lies at the center of this seventeenth-century chart showing the motions of the Sun, Moon, Mercury, Venus, Mars, Jupiter, and Saturn. For centuries astrologers based their predictions on this inaccurate and incomplete notion of the Solar System.

remained. Astrologers promptly invented a set of traits and influences for Uranus, rewrote their "ancient" systems to include the new planet, and stayed in business. They did the same thing when Neptune and Pluto were discovered, with no explanation of why those planets had never affected astrology before.

German chemist Martin Klaproth discovered a new element and named it uranium after the new planet.

An Explosive Honor

In 1789, a German chemist named Martin Klaproth discovered a new element, a heavy, radioactive substance. At first he named it klaprothium, but he had second thoughts and decided to call it uranium in honor of the planet that William Herschel had discovered in 1781. Many years later, the study of uranium would enable scientists to harness nuclear power, which became both an energy source and a deadly weapon.

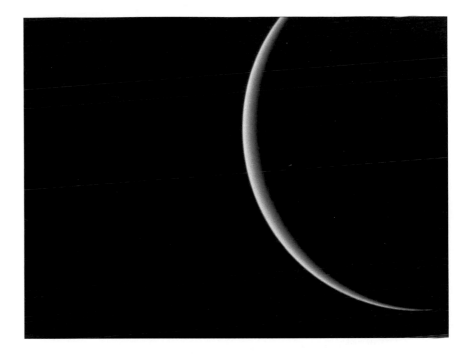

Uranus is nearly lost from view as Voyager 2 *heads away from the planet on its long journey out of the Solar System.*

FUTURE EXPLORATION

Voyager 2 spent less than one Earth day gathering close-up information on the Uranian system. Even so, scientists are still exploring the data it sent to Earth. That wealth of information, combined with new observations through telescopes on Earth and in orbit around Earth, may answer some of the questions that remain about the cloud-shrouded seventh planet.

For now, astronomers will have to be content with this possibility. There is unlikely to be another mission to Uranus any time soon. Although space scientists have grown increasingly interested in the outer Solar System, other projects have absorbed the money now available for space exploration. Neither NASA nor any other space agency has scheduled a second voyage to Uranus. Who knows? Maybe that will change before another Uranian year passes.

GLOSSARY

astronaut someone trained for space travel

astronomer one who studies space and the objects in it

celestial having to do with the sky, the heavens, or astronomy

constellation group of stars that appears to form a pattern, often named for a mythological being

diameter distance from one side of a circle or sphere to the other, through the center

geocentric describing a structure or system with Earth at its center

gravity force that holds matter together and draws lighter objects toward heavier ones

hemisphere half of a planet or other world, as in eastern/western or northern/southern hemispheres

interstellar having to do with the distances between the stars

nebulae clouds of stars and gases in space (singular is *nebula*)

orbit regular path followed by an object as it revolves around another object

probe   machine or tool sent to gather information and report it to the sender

satellite   object that revolves in orbit around a planet; natural satellites are called moons

sensor   instrument that can detect and record information, such as light waves, sounds, X rays, or gravitational and magnetic readings

Solar System   all bodies that revolve around or are influenced by the Sun, including planets, moons, asteroids, and comets

telescope   device that uses magnifying lenses, sometimes together with mirrors, to enlarge the image of something viewed through it

FIND OUT MORE

BOOKS FOR YOUNG READERS

Asimov, Isaac. *A Distant Puzzle: The Planet Uranus*. Revised edition. Milwaukee: Gareth Stevens, 1994.

Branley, Franklyn. *Uranus: The Seventh Planet*. New York: Crowell, 1988.

Brewer, Duncan. *The Outer Planets: Uranus, Neptune, Pluto*. Tarrytown, NY: Marshall Cavendish, 1993.

Brimner, Larry D. *Uranus*. Danbury, CT: Children's Press, 1999.

Kerrod, Robin. *Uranus, Neptune, and Pluto*. Minneapolis: Lerner, 2000.

Shepherd, Donna Walsh. *Uranus*. New York: Franklin Watts, 1994.

Simon, Seymour. *Uranus*. New York: Morrow Junior Books, 1990.

OTHER BOOKS

Armitage, Angus. *William Herschel*. London: Nelson, 1962. Part of the British Men of Science series, this biography covers the entire career of Herschel, who discovered Uranus.

Burgess, Eric. *Uranus and Neptune: The Distant Giants*. New York: Columbia University Press, 1988. Written after *Voyager 2*'s encounter with Uranus, this book summarizes the scientific knowledge about the planet.

Littmann, Mark. *Planets Beyond: Discovering the Outer Solar System.* New York: John Wiley & Sons, 1988. Much of this book deals with Uranus and *Voyager 2* flyby of that planet.

WEBSITES

These Internet sites offer information about Uranus and pictures of the planet, along with links to other sites:

www.bbc.co.uk/planets
Home page of the British Broadcasting Corporation's Planets site, companion to a television series. One section of the site is devoted to Uranus.

vraptor.jpl.nasa.gov/voyager/vgrur_fs.html
A fact sheet that summarizes the scientific knowledge of Uranus gained by the *Voyager* program.

pds.jpl.nasa.gov/planets
Home page of the Welcome to the Planets site maintained for NASA by the California Institute of Technology.

seds.lpl.arizona.edu/nineplanets/uranus/html
A Uranus site maintained by the Lunar and Planetary Laboratory of the University of Arizona.

www.solarviews.com/eng/uranus.htm
A collection of NASA images of Uranus and its system, processed for easier viewing.

nssdc.gsfc.nasa.gov/photo_gallery/photogallery-uranus.html
Information about Uranus compiled by the National Space Science Data Center, an arm of NASA.

ABOUT THE AUTHOR

Rebecca Stefoff, author of many books on scientific subjects for young readers, has been fascinated with space ever since her childhood when she spent summer nights lying on her lawn in Indiana, gazing up at the Milky Way. Her first telescope was a gift from her parents, who encouraged her interest in other worlds and in this one. Today she lives in Portland, Oregon, close to the clear skies and superb stargazing of eastern Oregon's deserts.

INDEX